MIND THE GAP

MIND THE GAP

Drystone Walling and Working with Shepherds in the Cumbrian Fells

Terry McCormick

To all the dry stone wallers who, over the centuries, have helped to make the Lake District a World Heritage Site.

Look out for the Stonechat! This is when I am lifting stones and chatting as I do…

When I first started drystone walling, I found myself chuckling as I recollected the London Underground command from my commuting days; 'Mind the Gap'. It was a private joke. Yes, that is what I am now doing in these Cumbria days.

My chuckles were checked just a little by the imperious and mournful tones of the 1970s 'Mind the Gap'. The premonition I could barely sense then was to become a force to reckon with as I worked alongside shepherds in the Cumbrian Fells.

Image on previous page: Mind the Rainbow, Tilberthwaite

Contents

Recruitment

'Look at this,' Julia said, sliding the Westmorland Gazette across the kitchen table. The page was open at *Employment Vacancies* and she had circled one which read: "Person wanted for drystone walling and shepherding work on a large hill farm near Kendal. Three days a week." 'Go on, try it,' she nudged, 'make the call.'

I was not at all hopeful when I picked up the phone.

The voice at the other end was warm, confident and sturdy. Had I had any experience before?

Yes, I had trained as a drystone waller and maintained our six-acre intake above Grasmere which still has gaps to mend.

There was a pause. This was of interest.

'We have 6000 acres of enclosed hill farm and we have a lot of gaps that need taking care of.'

I felt excitement stirring.

Then: 'I hope you don't mind if I ask you this question…' [*pause…well, yes I do because I think my fifty-five years will be a problem…*]. 'How old are you?' I give my age, thinking this conversation is about to end.

'Oh, that's fine, I was worried you might be in your seventies.'

I felt the soles of my feet settle into the floor.

'The next thing is to meet up I think. How about tomorrow? We are working in Bannisdale; do you know this place?' 'Yes, I do.' *Another pause…* 'I'll describe where we are.'

I know exactly where they are; the first gate along the track which marks the separation of enclosed in-bye fields and a boggy valley floor.

Bannisdale, where I was interviewed

What had seemed very unlikely ten minutes ago was becoming something tangible and possible.

There were three men around a wall gap, two with archetypal farming cloth caps, one, much younger, near a quadbike. The man I had spoken to on the phone came forward to shake my hand. This was Stephen. He introduced me to Brian and Dan. Me on one side of the wall, they on the other.

Rowan's Ground

Stephen gently questioned. What sort of walling had I done? Did I use a hammer?

Tiny nods of approval as I described the walls at Rowan's Ground; tough, gnarly, intractable stone not to be re-shaped easily. I said I was a fair waller, but I was slow. 'We don't mind slow as long as the job comes out well.'

Stephen had a longish, flat stone in his hands. He offered it across to me; 'Where would you put this?'

I looked nervously at the half-built course of stones. There was a nice space for it and it covered a join and added to evenness of the surface. I put it there and checked its stability. A half-nod followed.

'Well, we'll have to have a trial period and see how it goes. When can you come to Forest Hall Farm?'

Now I am relaxing and allowing a smile to broaden my face. 'Do I get to wear a cap like yours if I get through the trial?' This was greeted by chuckles by both Brian and Stephen; 'We might consider that, but it will take quite a while for it to become as distinguished as ours.' They were dead right on this. Both hats looked as if they had been around the world more than twice, with peaks badged with breaks and frazzle and seemingly held together by some sort of oily stuff.

Throughout this exchange and banter, Brian is mostly watchful, silent, and listening. Dan is in and out, vocal, present, but too young to show interest.

I walk back to the car. A fine day, clear and edgy. The fields brighter than when I arrived. A touch of feeling at home on this western boundary of the farm.

First Jobs

Posts up against the wall; metal wires threaded through chinks between stones; wire plier-twisted tight on the other side. Fencing rolled out and hammered onto a post and then stretched along to the next, tightened and tightened again with a ratcheter, secured to a post and then on, moving along the wall. Four of us working to prevent determined sheep from escaping. Most tasks need two pairs of hands; Brian doing the ratcheting, me clambering over the wall to fix the wire. We are along the track beyond High Borrowdale's now unoccupied last house. December days, early dusks, speeding the work against the cold and the fading light, chat mixed in with asks to hold, to twist, to stretch, to nail.

A request comes in to Stephen from a dealer who needs fifty lambs for a restaurant chain in Eastern Europe as quickly as possible, which means today. I head off with him in the Land Rover to the other farm he manages beyond Orton. The sheep are already penned up and we nudge and coax them into a paddock with a two-way gate. I follow Stephen's lead very closely. Moving sheep looks as if it should be simple; it is, if you are skilled and experienced and I am not. My job is to funnel to Stephen so he can assess and make his decision; for Eastern Europe or for staying at home. The day is bright and cold; the eight-month-old lambs understandably skittery. As the light catches their first-year coats, they become beautiful for a moment; fine-featured Swaledale faces. We work for almost three hours, holding sheep against legs before releasing, sometimes straddling; constant physical handling and effort which turns into struggle for this novice. As we work, Stephen talks about the sheep trade; the market system; how contacts work; what reputation is and what it stands for.

On the way back, he quizzes me about my background. I tell him of my family's small-farm way of life in the hills of mid-Wales stretching back into the early nineteenth century. He nods and affirms, 'Ah, yes, it's in your blood then.'

Forest Hall Farm is high on the expansive south-facing slopes of Shap Fell; Scotland forty miles to the north, Blackpool Tower, seen on a clear day, forty miles to the south. The fields on this slope curve away eastwards down towards the river. It is late afternoon, and I am tidying up some slips and falls in the walls. It is almost time to stop as the light ebbs away. I am beneath the huge pylons which have been constructed just outside the national park boundary. The wires are humming and sometimes cracking and snapping. Suddenly this huge rolling space of dark green and sky seems completely ancient and eerily relaxing. I am so tiny in the story of this land and as I gauge this, my attention is opening and opening; what a relief!

A Test

What am I really like as a waller?

This question is still there after a few days of settling in.

There is a substantial gap to the south of the farmyard, and Brian and I meet to work together there, him on one side, me on the other.

I make sure the wall is stripped out to where it is as solid as can be, sorting the stone as I do; rough filler, course stones from larger to smaller, throughs, flats, get-out-of-jail pieces, and cams rowed at the back. A good clear space for moving feet next to the wall. As I sift and set out, I worry that I am too slow. Brian's face and eyes are completely neutral. We begin to place and match, finding spaces, hardly speaking in a chuntering sort

(above) Test walling completed

(right and far right) Sequence of images showing a wall restored looking into High Borrowdale

of silence. This is where hands and eyes do all the talking and As I make decisions I am on total alert to what Brian is doing. Occasionally he has a stone which is just right for my side, and I return the favour, though less often. Courses in, joints crossed, filler thoroughly packed, the weight and locking-in of throughs shared.

Two men knitting stones in silence; just the clink and purl of faces and sides meeting, then joining.

The wall goes up steadily. The gap disappears. We both use our cams at the same angle and height; tiny compromises signalled with an out-breath. Brian's cap is nodding more frequently as the end comes into sight.

I'll do.

I am OK.

Brian tells Stephen I can wall by myself. I feel set free and grown up in a way I have never felt with other work I have done. I hum and sing quietly as I pack up for going home.

Scanning

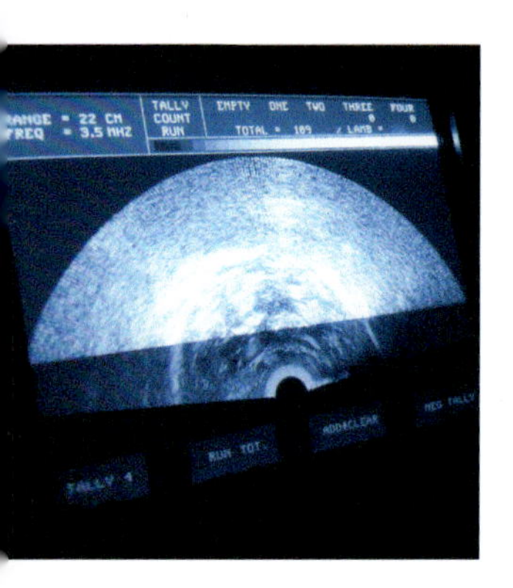

Looking at the future

It is about ten weeks since the tup was put to the ewes, and it is time to look closely at the results. The scanning man and his machine are booked in for evening sessions, and the sheep shed is turned into a waiting room for vocal ewes. My job is to encourage and channel the sheep to the screen so their wombs can be visually checked. Is it a single? Is it a twin? Is it a three-er? Or is it barren? Answers then influence feeding regimes and care until birth. Twins and certainly triplets will be indoors or kept close to the farm. The mountain environment at this time of the year, early January, will put too much stress on ewes trying to sustain more than one lamb.

I go outside the shed for a pee.

It is pitch-black apart from the light above the farm door.

As I turn back in, the other-worldly blue light of the scanning screen splashes onto the shepherds crouched in concentration amidst the creamy white flickering movement of the ewes.

For a moment it seems as if this technology has always been used in sheep sheds all over the fells.

I move to my place, tending, and join in this concentration on the blue hearth; a new womb, new unborn lambs, halfway to birth and a spring centuries away from this winter evening.

Ballast

There are two Borrowdales; the famous one south of Keswick and the much less well-known Borrowdale running from the River Lune at Tebay west into the remote eastern fells, bisected by the once great A6 northern route just below the Shap Pass.

West of the A6 is High Borrowdale, part of Forest Hall Farm, and today I am repairing a top wall in the last intake before the valley-head.

I park the car at High House. It is legs or quad bike beyond here, and I begin to walk, rucksack with bait and hammer, iron bar on my shoulder. I am in thick cloud mist.

High House almost in sight

(above left and right) Higher High Borrowdale on different days

As I follow the wall up and up to its right angle, I emerge above invisibility into bright early spring sunshine.

I set off with my routine, stripping, sorting, assessing the gap, clambering back and forth over the wall, selecting and levering in foundation stones. The cloud is thinning slowly as the sun gains strength.

I lean down, squinting to search across to the valley, taken aback by the bright green islands it has become, some with sheep grazing, one with a white farm house, and there, look, an island with a beck sending light signals.

I do not look too long or too hard.

Obscurely, I feel if I carry on with my task as if all this is routine, it will stay there. I crouch, bend, pick up stones and glance secretly at all that is below me. The effort and absorption seem to slow down time. The weight of rock is my ballast. With it in hand, I can sail between these islands.

A Bigger Test

I am to work alone on a serious gap in one of the lower fields beyond the copse to one side of the farm buildings. It is very damp ground, wet in places.

Once I have cleared out, I arrive at the foundation slabs. They seem fair. I stand on them and they do rock but rather than levering them out, I re-pack filler stones around them, thinking that the weight above will make them immoveable.

I work all day, building up almost to the top course.

Chaos cleared; foundations

Around mid-afternoon Stephen and Brian come along to see how I am getting on.

They look at the wall in unison and their attention immediately goes to the foundations. The wall above has bent and skewed the foundations into the damp ground beneath. I see it now with their eyes.

'This is not going to last,' Stephen says, and he and Brian set to taking the wall apart and asking me to help them.

I am in shock and a grim silence overtakes me.

When we get to the foundations, they lever them out, and begin digging down to firm ground, and then matching the underneath of the foundations with earth shapes so they settle without movement. Once set in, they jump up and down and where there is any movement at all they chock with small stones, and then re-fill so there is an even base to build on.

By now, it is almost dark; this is a short February day. I am exhausted and see all too clearly why this had to be done.

Stephen hands back the wall to me for tomorrow's work.

That night I think about short-cuts. In my life up until now I have thought of short-cuts as clever and to be used whenever possible. This had become engrained as a form of laziness. If I was to carry on walling, I had to leave short-cutting behind. When I woke up the next day, this short-cut reflex had gone.

April's Cruelty

My bedroom window in Kendal faces south and through the night had been battered by gales of rain, sleet and hailstones. From the womb-warmth of my duvet nest, I worried, projecting this up a thousand feet or so to the farm, exposed on the flank of Shap Fell with a typical 5+ degree temperature drop.

Against the walll, Tilberthwaite

There was no let-up in the morning as I drove up the A6, windscreen wipers barely managing. I kitted up in the shed and came into the yard.

Brian is standing in a storm that is just beginning to ebb, carrying what looks like a bag in each hand. I see his face first, the blood drained out of it. The bags are two dead lambs, their feet pinned together in each hand. He places them on a heap of tiny carcasses in the corner.

This is not a good day.

The fields to the south of the farm are usually good pasturage and the best area for lambs to find their feet with their mothers. Last night the usual rules were reversed as the most vicious weather hurtled in through 100 miles of open Lancashire and Cheshire plains.

The blizzards had swept twelve lambs up against a wall and, separated from their mothers, their fragile, new-dawn life was beaten and frozen out of them. Their bodies were now little sacs of just-formed bones, sodden, ice-cold, misshapen.

Brian is very upset and emotional. Not because of a loss of income, but because he is a shepherd and he is responsible for his sheep.

Heading home that evening and sitting in the comfort of Kendal's traffic jam, I reflected. This shepherding and farming with livestock have an emotional cost. Yes, most of the animals are destined for slaughter and consumption. It is a business. But underpinning this in the fell farms is the old knowledge that if land and livestock are thriving then so are family and community. Pastoral care requires a commitment that is not diminished by trading. In many respects shepherding is a deeply unsentimental form of love.

Clipping

We are in High Borrowdale, quads and trailers bumping along following the mobile clipping platform, landing up at a cluster of buildings backing onto the hill with a large paddock already full of about 200 very vocal Swaledales in their long June coats.

The two clipping lads (grown men in hill farming are 'lads') set up their gear, hooking into the precious electricity supply, a ramp at each end. We move hurdles around to set a throughway from a larger outbuilding to the ramp in what has become, for the day, a clipping shed. We divide thirty or so sheep from the paddock and encourage them into the outbuilding. My job is then to funnel (push and pull) them through to the clippers. This is a noisy job with clattery sheep spooked beyond cajoling.

The clippers are whirring away at top speed with swathes of wool falling to the ground as their confident cutting arcs produce short back and sides in two to three minutes. They work on a fixed rate per sheep and so there can be no dilly-dallying here. As one sheep jumps with relief off the platform, another has to be immediately presented. As I guide and sometimes wrestle sheep along, I begin to appreciate the strength and dexterity of these men (lads) producing shorn sheep with rarely a sight of blood; 'nary a nip'. The sweat is bucketing off them in the June heat.

At first, all this seems a long way from accounts of traditional pre-electricity clipping days before the 1960s. But the skill remains crucial with some clipping teams having such a strong reputation that they travel globally to do their work. And as I am not used to much companionship while working, comparisons don't come into it. The banter is pretty well non-stop and there is good crack at tea and bait breaks.

The clippers are part of a sub-contracting extended family network which takes care of farms with their very low manning levels. Forest Hall Farm has 6000 acres and there are two full-timers; me, and a manager dividing his time between this farm and another. Later in the year these lads will be at the farm's Christmas dinner.

The wool is piled onto a trailer and taken out into the yard where two of us spend the last hours of the afternoon packing (ramming) the wool into long wool sacks, secured with wooden pegs so simple that they must have been handed down through shepherding generations. Good work for chatting as you go along. There is sadness throughout, though, because the wool, so pure, so good, so beautiful, will fetch next to nothing at the Wool Marketing Board and clipping has to be done, and at considerable expense for the sake of sheep-health. Fifty or so years before and the wool-clip, it was said, paid a farm's annual rent.

The wool clip

Going Upwards

I push open the gate into High Borrowdale.

My destination is the top intake wall on the north-eastern flank of White How. The gap is up there in its far corner.

I set to, stripping and spreading, touching the stones I will place back. To touch and see makes memory; hand and eye will call out that stone when it is needed. The mugginess of the day shakes down as I get to the foundations, re-digging these in and jumping up and down on them to make sure there is no budging.

By midday the first courses are in, flat enough on top to invite another.

I sit and rest and eat my bait; slabs of cheese and pickle in fresh bread; a gourmet feast.

The wall ascends by my shoulder. In the middle ground the ancient pre-Roman trackway rolls over Shap, and just beyond this the A6 convoy of trucks and cars is too far away to be heard.

The silence of this traffic re-doubles the bird-song silence of the valley and so quietens me down that my breathing is at one with the rhythm of the breeze.

Without warning I am startled by happiness. I have been brought here for this. Tired muscles relax. There is a turning around, a revolving of all that I see into all that I am.

Bewildered and tender, I turn to my stone and begin to make the next course. I too am going upwards.

Foundations

It is Penrith Show day and I have entered the drystone walling competition.

Piles of stones of all shapes and sizes have been delivered and each competitor has to convert this into a completed length by mid-afternoon. This has to be done with a more or less permanent audience gathering to look and sometimes to offer advice and comments.

The walling is being judged by Steven Allen. Steven was a master-craftsman-god before being made famous by Andy Goldsworthy's commissions, one of which was in central park, New York. He was brought up on a farm near Tebay where he began his learning as a young lad helping his father.

We all set off, five competitors, mostly members of the Drystone Walling Association. I am nervous and going more slowly than anybody else, digging out shaped beds for carefully selected foundation slabs, wedging and earth-in-filling. My foundations lesson on the farm dictating terms. I finally get there, Steven keeping an eye on us all.

My section is going up, half a metre lower than the others.

A weary dejection begins to settle on me. I push on, trying to speed up and attract grumpy critical words from a fellow competitor because I am not using a guideline and relying on my eye. My farming practice – use your eye, be confident, move on, the next job is looming – was kicking in and this was not appreciated by somebody who was clearly right up there in the Drystone Walling Association hierarchy. I ignore him. It is too late for guidelines.

(above left) Walling on a steep bank

(above right) The finished wall and view across to Coniston Fells

Time is up and my wall is fair but definitely incomplete. By now I am rehearsing well-worn phrases of defeat; well, I turned up, I had a go, it wasn't my day, I can do better next time. Time for hot tea and plenty of cake. As Steven walks up and down the wall I look the other way, visualising the comfort of the farm's mountain fields.

The prizes are announced. I am joint third but, with a tiny shock, I have been awarded the prize for best foundations.

I am so delighted my hair begins to ripple. Steven gives me a crisp £5 note and a yellow rosette. My mood has switched from resignation to buoyant optimism. Fatigue has disappeared. The next hour or so is a blur as I help pack up the site.

When I get into the farm on Monday morning, Stephen and Brian are seriously impressed. This is a rosette for the farm and it goes up on the wall with all the others awarded for their stockman-ship skills with Swaledales. I do not recall any similar simple pleasure and pride in my life.

Castrating and Mothering

Holding a lamb is literally like holding a baby, but even more cuddly because of its soft wool coat. I bend down, collect a struggling lamb and snuggle my nose into its neck; mmmmm… this is a first and last new-born scent.

I then turn it over cradling its back in arms and hands, hind legs akimbo so that Brian can, with his clever device, loop a rubber band high up on its tiny ball sack. Within a few hours the loss of a blood supply will castrate.

Once back on the ground the lambs are understandably twitchy and at a loss. We keep them in the pen until there are a dozen or so before releasing them onto the fell slopes where their mothers, dazed and worried about their offspring, are making 'where is my lamb?' cries at full pelt.

We watch carefully as lamb and ewe noises are matched up and then confirmed by two-way nosing before frantic full-of-relief tugging and sucking at teats.

This can take half an hour or so, depending on distances between lambs and mothers. It is very calming to see them paired up and heading off together for mid-spring grazing.

Happy Toddler

I am holding tight on the trailer as we jig and swing into and out of tracks, weaving between peat hags and boggy ground on the way up. We have all the gear, iron bars, posts, tubs of nails and a deer gate, for a full day's work.

This day has to be chosen carefully; a clear-cut fine weather forecast is essential.

The watershed wall

Behind, the valley stretches away to Shap and the eastern Pennines, to the north, there are glimpses of the Carlisle plain.

Our destination is a watershed wall heading towards a T junction of valley heads. This is the outer edge of Forest Hall Farm. An ocean of hills on top of the world.

This wall is visited at most two or three times a year and there are many slips and falls. Brian and I work along quickly and practically, repairing and correcting.

Deer management for stalking is another strand to the farm's work, echoing the medieval function of much of this land as a deer park, known as Fawcett Forest. We are here to make it easier for the deer to get over the wall, putting in a moveable fence and making two cheek-ends for it; three good length stones into the wall and one good length across them, repeated.

The outer edges of Forest Hall Farm

To the west is Bannisdale Head where Brian was brought up before making the move across to High Borrowdale. It was a tradition for the two farmers from each valley to meet up at this wall once or twice a year to mend and make and enjoy a catch-up conversation.

As we work along and the hours go by, I feel as if a shell is being shed. Up here it is so elemental that all the accumulated junk of 'down there' is literally no longer needed, at least for a while. I find myself whistling and chortling in this lightness of being.

As I ride down in the trailer, swinging with the shape of the land, and looking out on the unfolding valley, I feel like a happy toddler, at last playing out in the places where I have always wanted to play.

(below left) Quad bike and trailer parked up for the day

(below right) It's grim up north

Khyber Pass

There are three Shap passes. One crowns the old road, west of the A6, which leads beyond Hollow Gate Farm where I park. Up there on the ridge to the north-west is a third U-shaped way over, or at least a Stone Age sight line for navigation. The team call this the Khyber Pass.

The wall runs parallel to this ridge and is stricken with gaping falls down to base-slabs which have twisted and buckled on wet ground. The foundations have to be completely removed and the ground dug down to firmness.

(below left and right) Before and halfway up below the Khyber Pass

I am not at all sure that I can do this.

Well, one action, one stone at a time. Let's get to work.

I dig out a bed to the width needed and then, levering up the slabs with my precious iron bar, set the uneven bumpy sides of the slabs in mirror shapes in the ground. Seven or eight of these I push and pull and iron-bar into place, with earth and stones rammed in any spaces left beneath. I do my stamping dance all over them, and then stand back. It looks sound. Good to go with the largest stones for the first course.

By lunchtime I am so tired that I am almost keeling over. Wrapped up against a chill easterly breeze, I sit and munch and drink my hot tea. It is going to be OK.

I muse contentedly amidst my aching bones. With iron came leverage on stone, and if I could do this then so much more is possible than our experience can comprehend.

I work on steadily that day and the next. Brian turns up on the quad to have a quick look, nodding before turning away. The wall is restored and strong and it represents my new-found confidence.

Whenever I come over this old road, I look up towards the Khyber Pass and then down towards the wall and register a reassuring nudge. I have done this. It will last. My confidence is made in the way this wall is made; enduring and interlocking with these fields and hills.

Shite Management (and a new gap)

So much of farming with livestock is about shite management.

My first job in the mornings is to clean up after the cows with the tractor and a drop-down six-foot-wide rubber squeegee; the pens, corridor between them, and the outside concrete yard.

And then, and this is the awkward bit, reversing and pushing the shite into the slurry pit. There is always a frisson of fear as I approach the edge, imagining a convulsion of gears and acceleration tipping the tractor and me into the depths of poo-oblivion.

But shite is magical as well as threatening. It feeds the land and makes pasture available for high nutrition grazing. It is essential

The bull

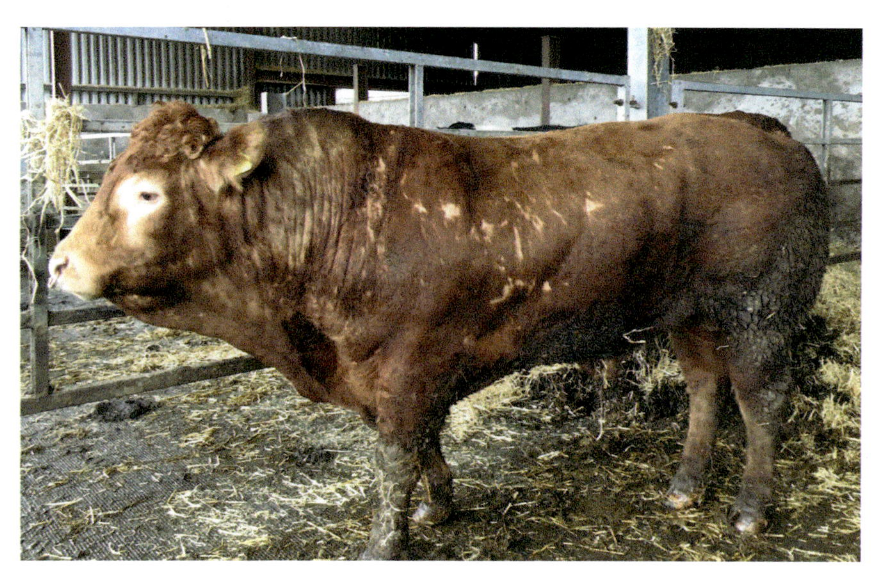

to the life of the farm and the land. I enjoy cleaning out the bull-pen by hand. There is a gate system so that as one space is cleared, the bull can be kept in another until completion. The solemn measured stare of this impressive Limousin seems too melancholic for a life of sex and endless rest.

One eye says 'Beware of what you want'.

And the other dictates and asks a question; 'I could easily squash you but why would I need to?' Both fix me with a gaze radiating out of some 1200 kilograms of potentially lethal force.

As the bull gets used to me there is a relaxation in his contemplation; 'Yes, you know your place'.

There is a small traditional cow-shed that can only be mucked out by hand. As I set to on what is a big job of work, I become obsessed with the challenge of it and dig and dig beyond fatigue into a sort of feverish frenzy. Some time, it is late afternoon, I hear inside me a distant muffled 'ping', as if a piano cord wrapped in wool has snapped. I carry on digging out beyond my physical capacity and must have called on my immune system to replace my body rather than allowing it to protect me.

A couple of weeks later I come down with shingles. My GP warns me; 'Beware; you are too old for this sort of work.'

Amidst all the gap restoration of walls on these hills I had managed, it seems, to open up a gap within my own inner field system.

My fragility is brought home during hay-making when I discover that I cannot get a bale above my head to give to the stacker. Luckily, one of the sub-contracting team sees my quandary and quietly and kindly moves alongside gesturing that I should give my bales to him so that he can do the upward chucking. My 55-year-old muscles are just not up to it but, more, I am struck by a total conviction that I have had it. My body cannot do this work.

Green, Dank, Damp and Cold

My most unpleasant gap yet: I am standing in a stream and suction bog in order to work.

Trees from January's storm have crashed onto the wall and, while the trunk has been cut away, the bole is up against the back which means I can only work on one side.

Stones have been scattered and buried and have to be probed out with the iron bar and chucked over.

I need to dig out the stream to get it to run in one runnel and then clear out under the wall and beyond it so that water does not seep into its foundations. This must have been happening for at least nine months.

'The river is rowdy with rain', Chapel Stile

Hard to figure this out. I will be stripping out and sorting for most of the day.

The wall is in the lee of an east-facing wood and is down a bank which prevents it getting any sun.

Green, dank, damp and cold.

Clouds white-out winter sun. The river is rowdy with rain. A wren scouts and forages in and out of the wreckage.

This wall is such a mental challenge. I am convinced that I will not have enough stone. There are an unusual number of throughs which I'll have to use carefully to ensure I get some height.

By day's end I have done so much better than I thought I would. The wall-section is at least a third up with a good set of throughs on top. The next day, order is made out of chaos.

Moleing

Who's in charge?

Molers are in demand. This is a job that farmers no longer have time to do. Brian took pride in High Borrowdale being molehill free. Now they are appearing everywhere.

Stephen thinks this could be a good complementary earner for me and proposes a learning session with Brian.

I am nervous. I like moles. I like that they seem blind but can find their way; that they are such powerful diggers; their faces are all nose with excavators on either side. Their single-minded work rate and their survival underground are noble. And then, when I hold one, they are so soft and velvety.

We have an afternoon in a hill-riddled field, probing earth to find their underground motorways, and setting traps in these and covering them up.

We find one, still alive, with the trap lever crushing its eyes.

I am not tough enough to be a moler.

I reflect later that I can never be a proper farmer. Molehills reduce grazing and they are a highly visual reminder that the farm is not on top of its game. They need to be dealt with.

Not by me.

The Point

It is January and the temperatures have plummeted. It is raining ice (or is it glass?) on an easterly gale.

I have decided to work on this stretch of wall so I push on with my Siberian headgear buttoned tight.

There is a point to prove and it is stuck somewhere along an edge which I have to traverse, apparently, if I am to know who I am.

Am I tough enough to make and build in the jaws of this hostility?

Yes, but not without a price as I crouch and lift stones, knocked breathless by this cold.

I lose.

There is no further to go.

I stagger back to the protection of the car, numb and baffled at my stubbornness.

That night I wake up and decide, too late, that I no longer need to impress myself or anybody else.

Frozen jaws

Around the Heart

On this evening, my usual post-supper pottering around the house has become twitchy and I am unsettled; and then feverish, and then nauseous.

I don't feel well. What is going on?

Suddenly, I keel over with extreme and crushing chest pain.

I am choking on words I am trying to make.

Ju rings 999.

Within ten minutes I see NHS legs, green against the russet of our sofa. The walls are strobing silver-blue.

Paramedics are asking questions. Where is the pain located? How long has it been here?

They tell me we are going down to the Cardiac Unit at Westmorland General Hospital.

I am flitting in and out of blankness.

As we speed down the hill, wires are attached and information is phoned ahead to the unit ten minutes away.

I deny that I am this body and float up above the ambulance floor, puzzled and bewildered. Is that really me? No, it can't be.

I am bumped back into my self by white hospital light as I am scooted along to the unit where everybody seems to be ready, poised for action.

Through the fog of pain-reducing medication it becomes clear within an hour that this is not a heart attack. My helpers are baffled and relieved.

I do not know what I think or feel. The clock tells me it has just gone 10.00pm.

This is a heart

I stay in the hospital all night, closely and regularly checked. I am sick. Then, halfway through the night I have an acute diarrhoea episode. Though still feverish, the pain is reducing and slowly ebbing away.

I am not dead.

In the morning the consultant visits and I learn that I have had something called pericarditis. The two thin layers of tissue that surround my heart and hold it in place have become inflamed and they have been rubbing against each other and my heart. The pain is very similar to a heart attack.

The most likely cause is a viral infection.

There is no treatment. This is a random event which is unlikely to be repeated.

I am sent home and told that I must rest and not worry.

The Banal versus the Profound

Standing beneath the arch of a rainbow, one end descending on the trees above the gap that I repaired last week. I step back from the wall to watch and wonder; another rainbow comes out of my recollection and joins the one I am looking at now.

Frequently, when I build, I cease to exist; I'm either dreaming or wide awake; but whichever, I am not here and there is no time anywhere. When I 'come back' (usually because of fatigue or hunger or both), I know that I have been gone. I look at the work I've done and see myself there.

Walling all day alone, the banal jostles with the profound and, towards the end of the day, seems to be winning. I'm sitting in the sun, back to a sheltering wall, looking eastwards at the long and high Shap Pass. Nothing is on this usually busy A6; it is empty upwards and downwards. As I munch my lunch, a thinking bubble floats out the question: 'Could this be the end of the world? And if it is, what shall I do?'

This houmous and grated carrot sandwich is so delicious. The streamlet gurgling at my feet is a soothing music inside my veins. That wind whoooshhing up the scrub and fell-grass is cleaning out my lungs.

I don't need to do anything because 'I' am not here...

As soon as this is a thought it is immediately banal and the trucks and cars become noisy as they drag up the A6.

'One end descending on the trees', Loughrigg Fell

Few but Absolute Rules

The rules for drystone walling are few and absolute; ALWAYS:

Strip out the decay quickly and completely.

Arrange the stones in groups; filler for inside; big, heavy and ugly for foundations; awkward stones that are too big for filler but will often get left until last; cam stones rowed at the back; bigger course stones nearer the builder; smaller further away; good-looking flats; special 'get-me-out-of-prison stones'; throughs spread out to the side

Place the flat side upwards and strive for level courses (for foundations this means digging out a shaped hole for the stone to fit into)

Cross the joins (only possible if the courses are as flat as can be)

Place stones lengthways-in even if this means sacrificing a fine face

Make sure each stone is left secure and not rocking or moving

Use a string if the wall is more than six metres in length

Decide on which courses to place the throughs (depends on how many throughs are available)

Place throughs on joins and not above each other

Use chunky bouldery stones to match the throughs on that course

Make sure that good stones are not lost inside the wall

Use the hammer sparingly

Think of patterning; two courses of chunky, one of thins, depending on pre-existing wall

Prefer strength to beauty, but strive for both

The stone is ALWAYS innocent; do not blame it.

Wriggling outgang, on the way up to Alcock Tarn

Swaledale Ewes

(Almost a sonnet)

In all points they are high scorers:
elegant grey-white jaws beneath
black shapely faces and yellow-ring eyes;
legs designed for a standing start,
wool made for northern weather.
Character and bodies are at one.
She is pure Mother, super aware,
alert to all dangers, shapely ears
on-the-listen for her purposes.
Sensitive and skittish, she looks
out for her brood with constancy;
lambs which tug the toughest heart-strings.
In a Mrs World Mother competition,
she is always a finalist and usually a winner.

'Strength to beauty',
Little Langdale

Rant

The wall gap is alongside the old packhorse track that once was the main route over Shap. Fallen layers making a frame for a fine view southwards. I am about to delete this.

I like the title for these recollections because of its humility. No grand purposes or missionary zeal here. Landscape does not need these, but, rather, endless acts of care and restoration by as many people as possible, and most often by farming families. When the gaps are minded the unity of walls and their places is re-set and, visually, the hills are settled and reassuring.

The landscape framed

I became aware that 'gapping' was looked down on by many hobby drystone wallers who were aiming to be craftsmen with a

capital 'C'. Gapping is seen as the most basic form of walling skill, to be learnt and then to be left behind. Farm wallers are regarded as too pragmatic, too 'make-do'. I have heard the description of a farm wall being 'thrown up' more than once.

But hill-farming families most often do a great job keeping walls mended and neat in circumstances where the stone is as found, often resistant to a hammer, even if there was time enough to shape it.

Walling amongst the fells is frequently awkward, often on slopes where the stones cannot be spread out, enforcing frequent clambering up and down climbs with stone in hands and arms. These walls do not receive attention and are left to farmers and shepherds. But, of course, the very best wallers have often been apprenticed on 'gapping' from an early age. The showpiece completed wall from foundations to copes is the default image holding full page, as if this is more important than a mountain wall made complete by restoration.

The landscape wall-settled

The Lake District's drystone walls are the nation's equivalent of the Great Wall of China and if all the leaders of the key agencies (they know who they are) committed, like a good farm or estate, to having no gaps, that would be a tangible and highly visual salute to World Heritage status. How immortal would National Park officers be if they did this rather than processing their processes in their processed offices.

Today, these are my thoughts, this is my rant.

I am not far from what must be the highest farm in England. The Knowles family have been taking care of Rough Fell sheep here for many decades. This is Rough Fell country stretching eastwards towards Sedbergh.

If Herdwicks are the most charismatic sheep, and Swaledales are the most glamorous, Rough Fells are the best dressed, in gorgeous fleeces, and all carried off with heads and jaws that would make Churchill envious.

As I work, Tommy Knowles, an 80+ year-old 'retiree' is doing his quad-bike round checking up on the sheep. I see him in the morning, and we both comment on the weather. By mid-afternoon the gap has almost disappeared, and we pause for a longer chat. As he goes, he looks back and says; 'It's a tidy job that.'

This is praise indeed from somebody with so much experience and I am buoyed up.

We are as good as our work. All distinctions are made from this core world view, and what a relief and release it is from a pervasive contemporary culture of status, money, celebrity; soft centres of smugness.

The last course and then the cams are positioned and I stand back and affirm what I see before tidying up and relaxing into the front seat of the car with a sigh of relief. The job has been done and I am pleased with its tidiness.

Great walls of the Lake District

Meat

Stephen, Brian and Dan do such a brilliant job taking care of these sheep, and especially the growing lambs. Any lameness, infection or disease is dealt with immediately and prevention rules. They look so bright-eyed and healthy; if they were pets, their owners might be accused of molly-coddling.

The farm has been organic since 2001 which, with pasture without inputs, means special attention to pharmacy and feed. When I am bagging and sharing out the food granules, I know this stuff is good. I could put this in with my breakfast muesli.

We are all chuntering away while moving lambs into the shed and the conversation turns to food. I own up to being a vegetarian. There is immediate shock and bafflement all round. What was I doing here if I didn't eat meat?

Stephen leads the questioning; why and how? I say if I could take care of an animal the way they do and then kill and butcher it myself, I would happily eat the meat. Stephen gets this straight away and says he will do that with me one day.

Slaughtering surrounds this discussion. Dan does not like to think of what happens to the sheep when they leave the farm.

Later, I learn that European Union regulations following the BSE (Bovine Spongiform Enchephalopathy) calamity have entailed huge industrial slaughterhouses with resident health and safety inspectors. This is the only way it can be done economically.

Sheep and lambs from Cumbria travel as far as Somerset to be slaughtered.

Smell

How can I describe the smell of a lamb?

Baby brand-new folds of soft skin.
Mother feeding.
Sweetness and sharpness of cut-grass.
Spores of wool-scent; lanolin.

Aromas of fleece infuse clothes, skin and hair.

Handling sheep all day means I smell like them.
It doesn't simply wash away.
It has become my scent.

I like this smell.
It is earthy, medicinal and ancient.

All shepherds since time began will have this bouquet on and around them.

So much about lanolin is balmy and healing.
It is unavoidably good.
It keeps sheep dry through rainy days and rainy days.

When I get home, I have to leave my clothes in a pile outside the house or my wife will smell like me.

At the farm Christmas party, the tale is told of this scent surviving hot showers and getting into a glass of wine.

It cannot be escaped.

Prospective partners of shepherds, beware.

Knowledge

This is a knowledge that does not require technology or much pen and paper.

Brian was brought up to 'ken' sheep, looking at a flock and knowing each sheep individually without numbers and tags.

His ability to move amongst livestock without unsettling them is uncanny; skittery Swaledales and nervous beef cows remain serene in his presence. He knows how to be absent; like a hunter but in a caring role.

He can scan a field and tell you where the moles are travelling.

Brian knows High Borrowdale better than the back of his hand. He has lived there since he was ten and so all the sheltering places for sheep in storm or summer heat are known to him.

As he says, tongue in cheek; 'You can get it all on t'internet.'

He believes that traditional shepherding is traditional because it works. It was worth keeping and handing on. He does not approve of lambing in sheds to get an early crop because hill sheep need to be hardened by the hill environment. Sheep diseases require pharmacy, yes, but disease is also a way of strengthening the flock's bloodlines. Imported foodstuffs are a luxury to be tolerated. He recalls many times when they could not be afforded.

Brian's physical capacity for a 60+ year-old is impressive. I have seen him chase a ram, do a flying tackle to grab its horns, flipping it on its side, rodeo style. There is no doubt who is boss.

When I was working with him, the shadow of foot-and-mouth disease was still present. Forest Hall Farm had been culled and the flock was being rebuilt using organic principles. But this was a watershed calamity. In 2001 he walked into High Borrowdale and its absolute silence and sheep-emptiness after centuries and his own lifetime overwhelmed him. The psychological cost is high, and he reads this event as a sign that the story is coming to an end.

With very few words Brian shows the way with walling. He has an inner relaxation with the craft of placement which, on the hill, with all choices as found, makes stone selection and compromise an art form. Over time I begin to try and mirror his approach. My learning is a day-by-day imbibing through watching and mimicry.

High Borrowdale Farm just and so in sight

The Boss

So much of my working life had been bedevilled by boss problems and office politics. I know this is an experience shared by many.

Stephen, my boss at Forest Hall Farm, is an impressive counter-punch to all this.

He is a man comfortably at one with what he does. The consequence is quality and achievement; he is definitely a leading member of the A team.

The man-management seems effortless with complete equality in the sharing of work while always linking into a perpetual motion big picture: the life of the farm.

His voice does so much work for him; persistently affable and warm.

His eye for a good sheep is renowned and reflected in numerous show rosettes. The determination is there in the pre-show pampering that we all do for his Swaledales before a competition. They are made pure creamy-white and luxurious by oodles of fairy liquid and warm water. Bottoms are as perfect as a bottom can be. This is serious sheep-glam.

Whatever stereotypes of hill farmers there are, Stephen bucks them. He is a historian of his landscape, puzzling over Bronze Age settlements and more recent tracings. One of his holidays is in Rome. He tunes into what is going on in the world of politics, and keeps his eye on the share-price.

If Stephen wasn't so good at what he does, he could easily be a college lecturer. As we work along, walkers sometimes pause and try to figure out what is going on, then, hesitantly, ask a question.

The best repairs are invisible, almost...

He explains clearly, knowing where they are coming from and leading them into the work, his own enthusiasm shared and enjoyed.

Stephen thinks long and reflectively about fell farming.

If there are prejudices he keeps them to himself; as three of us squeeze up close in the front seat of the Land Rover, he quips; 'This will get the neighbours talking.'

For me he is a gift. What you see is what you get and I am able to flourish.

There is time in the second season when he and I are mothering up the sheep, and he lets me know that I have learnt to do most jobs twice and I am definitely underway. It feels a bit like a graduation moment and I cherish it.

On Being Blue

As I arrive at the top wall separating pasture from woods, my eyes gasp at the shock of this blue.

It is mid-May and an early morning. Columns of light are rising out of roods of blue bells. Here and there is a patch of green and bracken fronds pelmet the wall.

I am on the shore of a bluebell sea sorting and spreading loose, friable, slatey stone. The waves lap against a hillside and as I look up I see slopes reflecting bright sunshine.

Something is tugging downwards in me, as if I half-understand that this is my last time. The procession of minutes is a sort of aching, shot through with an incessant blue yearning to be ever-present. I am leaning at an open window looking out of time.

Déjà blue

As the wall goes up I become a converter and restorer absorbed in hand's memory as my body follows the sequence whatever the weight.

I am on the edge of this bluebell wood and on the edge of an old self; all the elements catch at my feet and make my eyes feel rather than see their way.

The whole of this blue world is seeing through and with me. I breath out quietly and rest in an immense relief at no longer being trapped in importance and fretfulness. The stone is such an old story and I have been here before in this moment.

The déjà-vu settles me: I sit down.

Time for a break.

I pour out the tea and raise my mug to this ocean of everlasting blue.

Suddenly in the Side

I have another 'suddenly' event.

Waking early, I am unable to breathe. Well, I can, but only with stabbing pain, especially in my left side.

Moving up and down is agony too. I establish one position where it is bearable and I lie fixed in this with my mind Ferris-wheeling questions and questions.

The GP has to be brought to me. Pleurisy is diagnosed, a viral infection blamed, and a course of antibiotics and anti-inflammatories prescribed.

This will have to be endured. It will cease at some point.

I am laid low for just over two weeks.

Pleurisy, I later learn, is sister or brother to pericarditis. Two layers of tissue, one lining the inner chest, one surrounding the lungs, are inflamed and irritated, and then sandpaper each other.

I have hours to mull over the mystery; shingles; pericarditis; pleurisy; the increasing grip of fatigue; and, recently, weight loss.

My body is the host.

Is it a Wolf at the Door? (I'm not chuckling now)

It is Tangwyn's graduation day and we all gather in Cambridge to enjoy it and celebrate with him.

I am too excited. Adrenaline takes over as we walk from one end of the city to the other through a long summer day. By the evening I am immobile and helpless with tiredness.

At home, my knees and ankles swell and flare as if they have crashed into a wall at speed; I can hardly walk. I fall into bed once more.

The gap here which cannot be minded is between what is going on and a complete inability to understand.

I am a stranger in the foreign land of my own body. Hormones are mixed up with feelings. Nerve endings are numb. The exhaustion is aggressive. And now, dark red rashes and blotches decorate ankles and shins.

Julia has a colleague who suffers from lupus, and knowing how this illness travels around the body searching for a home, she wonders aloud.

I share her thoughts with the GP and I see an ah-ah light-up moment in his eyes. He doubts lupus, but he says we must now look at everything as if it is connected.

I am signed up for every ultra-test there is and tagged 'urgent'.

I hold onto the word 'connected'.

It becomes my conviction, still weighted with unknowing, but a breakthrough.

Mind the Gap

It is now definite. I cannot deny it. I am unwell.

Gaps have been opening in random places throughout my immune system. At first tiny and tolerable, and then bigger and intolerable.

I could not read the signs and nor could my doctors. I pushed and pushed on, with a fiendish logic that dictated: 'If you can do this then you must be OK. Go on; prove it.'

Stephen, taking in my enforced absences, evident weight loss and increasing pallor, asks for a pow-wow. We agree that I will stop work until I know what is going on and am getting better.

These are the last of my drystone walling days on the farm and the fells.

At this time, I had also taken on a commission to develop a cyclothem wall at Killhope Lead Mining Museum over on the North Pennines Watershed.

I was excited to have this challenge but apprehensive because of this steady fall away of energy. Luckily, I was able to commission a skilled local drystone waller, Anthony Page, to get the wall standing.

The story of this wall matches the story of my illness and rehabilitation.

A cyclothem is the internal geology and hidden shape of a landscape. The objective of the wall is to embody this through stone selection and patterning, and to reflect the external shape of this landscape as well.

(opposite top left] Hard shale in quarry

(opposite top right and below) Cyclothem wall

The internal formation of my own body's landscape demanded more and more attention as it went through its slow-motion collapse. The cyclothem wall echoes this because of the challenge of shale.

Shale needs to be represented as four layers within the wall in the following order, from on top downwards:

Firestone Sill (sandstone)
Shale
Little Limestone
Shale
Great Limestone
Shale
Quarry Hazle (sandstone)
Shale
Four Fathom Limestone
Natrass Gill Hazle (sandstone)

Shale is friable and in a constant state of decomposition and would not, therefore, provide a course that would endure. A drystone wall with shale so dominant would be unstable, stirring up disruptive, random fissures and impacts.

This wall and my immune system could be twins.

I am diagnosed with a rare version of something called vasculitis. The immune system's innate creativity, I learn, has produced a new cell, an antibody, as part of a kind of blind coping. These antibodies have been steadily taking over for many months. Like the shale in my cyclothem wall, they cause decomposition in the structure of cells and tiny blood vessels.

Finally, now, my kidneys are a favoured home for these antibodies. They have moved in to stay. By the time the situation is grasped, my kidney function is close to 30% and I am days away from dialysis.

This is what the antibody looks like under a powerful microscope.

The foundations are all over the place. I do not know if they can be gathered and re-set again.

The eventual solution for the cyclothem wall's decomposing shale is to import a hard shale from a quarry near Ambleside.

This is then soaked in a very dark linseed oil so that it takes on the coal tones of friable shale.

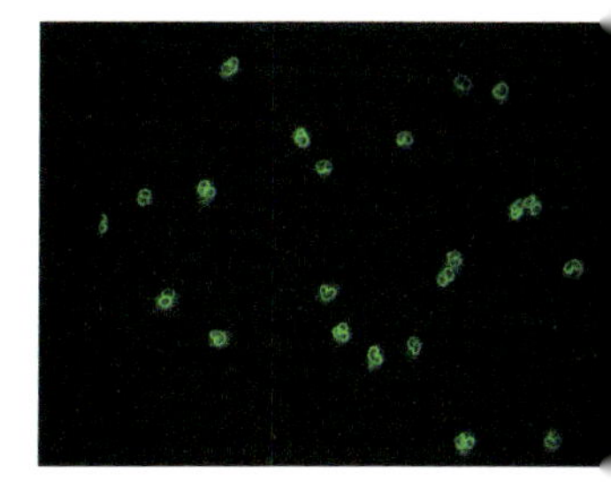

Antibody p.anca, later known as 'Pete Anca'

Solutions for my body's perishing were going to be more intricate and more dependent upon skilled help from physicians and specialists.

A first step is to replace all my toxic blood with synthetic plasma. My new blood is the first foundation being put in place.

When I had started drystone walling almost seventeen years ago, I had chuckled as I recollected 'Mind the Gap' on the London Underground.

Now, the voice came back to me as a premonition I could never have appreciated at the time.

I had not followed these instructions with sufficient awareness. A gap had been made where there shouldn't be one and I fell through it.

My chuckles stop for a while.

Then they are gradually nudged back by the exhilaration of survival.

Over three years, while going through my own treatment programme, I oversee the building of the cyclothem wall.

Its completion marks the beginning of a new life for me.

Acknowledgements and Thanks

My first thanks are to Stephen Lord for employing me at Forest Hall Farm and being my boss. Thanks too, to Brian Nevinson, shepherd, for being my co-worker and mentor. Both Stephen and Brian taught me so much.

Thanks to James and Helen Rebanks for stirring me to write about how it *felt* to work as a waller and shepherd's assistant. This got me going on *Mind the Gap*.

When I was taking photographs as I worked at Forest Hall Farm, I had no idea that this book would be written and produced. Many of my 'snaps' were just not up to it, and I am very thankful to Rosemary Everett, Rob Fraser, Wayne Hutchinson, and Julia McCormick for making up for this lack with their photography skills.

Dallin Chapman, Julian Heaton Cooper, Erwyd Howells, Elen Howells, James Macdonald Lockhart, Stephen Lord, David McCormick, Julia McCormick, Jim Troughton, Gary Wilson, Karina Wilson and Yasminca Wilson have read drafts of the text and made valuable comments on these and the photographs.

Grevel Lindop invited me to read parts of *Mind the Gap* for a Temenos lecture in London in September 2018, and facilitated useful feedback.

Sarah Stoddart invited me to read parts of *Mind the Gap* to members of her 'Sight Advice South Lakes' group in October 2019 who gave me much encouragement.

Thanks to Marianne Birkby for her drawing of a Stonechat on a camstone.

Thanks to Pat Sumner for proof reading the text.

Thanks to Charlotte Dean, Nick Fecit, Ian Forbes, Pru Kitching, Bridget Halldearn, Elizabeth Pickett and Brian Young for all their help in developing the cyclothem wall at Killhope Lead Mining Museum. Thanks especially to Anthony Wood for his drystone walling skill and patience. The cyclothem wall at Killhope was commissioned by the North Pennines AONB Partnership and supported with funds from Natural England, County Durham Environmental Trust, and the Heritage Lottery Fund.

Thanks to Isobel Gillan for shepherding me through this production with her book-design skills and patience.

Photographic Credits

Published in 2021

Designer: Isobel Gillan
www.isobelgillanbookdesign.co.uk

Other publications by the author:

'Wordsworth & Shepherds' in *The Oxford Handbook of William Wordsworth* (2015)

Lake District Fell Farming: Historical and Literary Perspectives, 1750-2017 (2018)